STUDENT GUIDE

LOOKING BEHIND THE NUMBERS

CORRELATIONS, RANKINGS, AND PERMUTATIONS

MathScape
SEEING AND THINKING
MATHEMATICALLY

How can
math help
you describe
and analyze
data?

LOOKING
BEHIND THE
NUMBERS

PHASE**ONE**
Statistical Measures

In this phase you will be collecting and analyzing sports data. You will find means, medians, modes, and ranges. You will also see how stem-and-leaf plots can help you display a lot of information very compactly. All of these tools will help you to make fair comparisons.

PHASE**TWO**
Data, Scatter Plots, and Correlations

You will begin by making some measurements of your own body. Then you will have a chance to use this data throughout the phase. You will see how to use scatter plots, lines of best fit, and correlations to describe relationships in data. At the end of the phase, you will use everything you have learned to help solve a mystery.

PHASE**THREE**
Probability, Combinations, and Permutations

Imagine that you and your classmates have formed a musical group. You will have many choices to make about concert locations and song sequences. The investigations in this phase will help you use combinations, permutations, and probability in answering these types of questions.

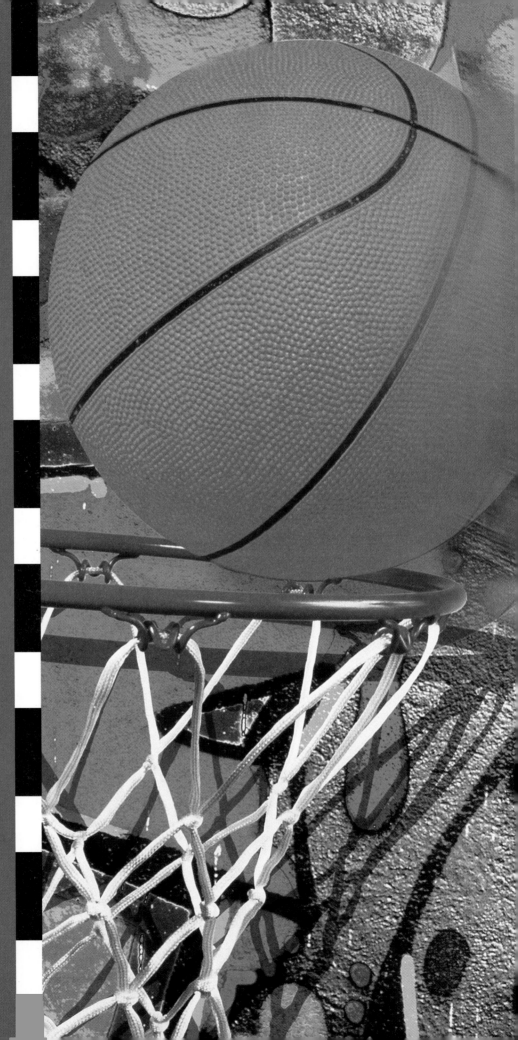

PHASE **ONE**

In this phase, you will be collecting and analyzing sports data. By using means, medians, modes, and ranges, you will be able to compare the performances of individuals and teams. Stem-and-leaf plots will give you a powerful way of displaying data.

Batting averages are just one example of how data and statistics are used in sports. What other examples of sports statistics can you think of?

Statistical Measures

WHAT'S THE MATH?

Investigations in this section focus on:

STATISTICS: DATA COLLECTION and REPRESENTATION

- Collecting and organizing data
- Making stem-and-leaf plots

STATISTICS: DATA ANALYSIS

- Calculating means, medians, modes, and ranges
- Finding data sets that fit a given mean, median, mode, or range
- Making conclusions that are based on data

NUMBER

- Using percentages to describe data

MathScape Online
mathscape3.com/self_check_quiz

1 Sports Opinions and Facts

Surveys are useful tools for collecting data about people's opinions. You will take a survey about sports and use averages and percentages to analyze the class data. Then you will interpret tables and graphs to find out the facts about in-line skating injuries.

Collect and Analyze the Class Data

How can you use means and percents to analyze data?

The class will take a survey to find out students' opinions about in-line skating. Then analyze the data by finding the following:

1 How many students chose each rating?

2 What percent of the students in the class gave each rating?

3 Add up all the ratings to find the total rating.

4 What is the class average, or mean rating? To find the average, add up all the rating values that students selected and then divide by the number of students who gave these ratings.

Sports Opinions Survey

Rate in-line skating on these scales. Write your ratings on small pieces of paper.

Rating Scale A: How much fun do you think in-line skating (rollerblading) is?

1	2	3	4	5
No fun at all		OK		Great fun

Rating Scale B: How likely do you think it is for people to get injured when they go in-line skating?

1	2	3	4	5
Very unlikely to get injuries				Very likely to get injuries

Interpret Graphs and Tables

The handout In-Line Skating Data provides information on in-line skating injuries. You might be surprised by what you find out. For each question, write at least one sentence and give specific data to support your answer.

1 Did more injuries happen outdoors or indoors?

2 Were more injuries on arms or legs?

3 How much experience did the injured in-line skaters have skating?

4 Did more injuries happen to people who didn't take lessons or to those who took a lot of lessons?

5 What percent of the people injured were wearing all the recommended safety gear? What percent were wearing no safety gear?

6 How do the number of injuries compare for in-line skating and gymnastics? Why do you think this is the case?

How can you interpret data from multiple graphs to get an in-depth picture of a topic?

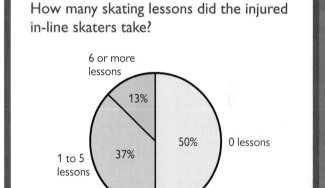

Count the Lessons

How many skating lessons did the injured in-line skaters take?

6 or more lessons — 13%

1 to 5 lessons — 37%

50% — 0 lessons

Write Questions About Data

- Write two new questions about in-line skating that you could answer by using data from the graphs and table.

- Write two questions on in-line skating that you would like to find out about but do not have the data to answer.

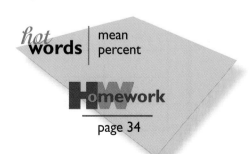

*hot*words | mean percent

HW**omework**

page 34

2 Who's the Best?

Means, medians, modes, and ranges are statistical measures that are useful for figuring out what's typical and what's unusual. How can you use these measures to determine which basketball player is the best?

Determine Means, Medians, Modes, and Ranges

How can you use means, medians, modes, and ranges to analyze athletic performance?

The first table of Basketball Data shows how many points three basketball players scored in seven different games.

1 Look at the example for Player A in the second table. Then make a copy of the table and figure out the missing statistical measures for Players B and C.

2 After you complete the second table, use these clues to help you figure out which player is which. The three players' names are Sheryl, Neisa, and Tina. Sheryl's median score is the same number as Neisa's mean score.

Basketball Data

Game	Player A's Points	Player B's Points	Player C's Points
1	12	18	24
2	13	21	14
3	12	15	14
4	14	13	22
5	11	16	25
6	20	18	16
7	16	18	11

	Player A	Player B	Player C
High Score	20	?	?
Median	13	?	?
Mean	14	?	?
Mode	12	?	?
Range	9	?	?

Compare and Rank the Players

If you compare the players' high scores, then Player C is the best because 25 is the highest score. Will Player C continue to be the "best" when you use the other statistical measures? Figure out the other rankings and put them in a table like the one below.

Do different statistical measures lead to different rankings of who's the best?

Different Ways to Rank the Basketball Players

	Ranked by High Score	Ranked by Median Score	Ranked by Mean Score	Ranked by Mode	Ranked by Range
Highest	C, 25				
2nd	B, 21				
3rd	A, 20				

After you complete the table, decide which player or players you think are best overall. Who do you think is second best? third best? Be prepared to explain your conclusions.

Write Conclusions

The basketball coach is planning to give an award to the best player, but she can't decide which of the three players is the best. She has asked your class for help.

Write a letter to the coach to convince her of one of the following arguments:

- that one of the players is the best
- that two of the players are the best
- that all the players are equally good
- that it doesn't make sense to give out this kind of award

Make sure to use the data to support your conclusions.

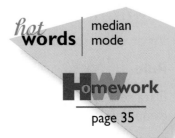

hot **words** | median mode

Homework
page 35

3 Mystery Data

If you know that a basketball player's mean score is 10 points, what might each of his or her individual scores have been? Working backwards will help you build a better understanding of means, medians, modes, and ranges.

Solve Clues About Data

How can you create data sets that have specific statistical measures?

You and a partner will play a Mystery Data game like the one below. For each game, use a calculator to help you find a data set that matches all the clues. Write your solution with the values in order from least to greatest. Be prepared to prove to the class that your solutions work.

Sample Clues for a Mystery Data Game

A. Tanya played 5 basketball games and kept track of how many points she scored in each one.

B. Her mean score was 8 points.

C. She never scored 0 points.

What might each of her scores be? Find a data set to match all the clues.

$\underline{?}$, $\underline{?}$, $\underline{?}$, $\underline{?}$, $\underline{?}$

Sample Solutions

6, 7, 7, 7, 13   5, 7, 8, 9, 11

What other solutions can you find?

Rule

Data sets like 8, 8, 8, 8, 8 are not allowed in the game because they are too easy to find.

Experiment with Means, Medians, and Modes

Seven students kept track of how many hours they played sports during August. Their mean was 21 hours. How many hours might each of the students have played sports? Your challenge is to make data sets to match the descriptions below.

Choose four of the descriptions and make a different data set to match each one. All the data sets should have 7 values and a mean of 21. In each of your data sets, circle the median and put a square around the mode (if any).

A. Mean is larger than median.
B. Median is larger than mean.
C. Mean is larger than mode.
D. Mode is larger than mean.
E. Median is larger than mode.
F. Mode is larger than median.
G. Mean, median, and mode are equal.

How can you create data sets that have different relationships among means, medians, and modes?

Write Your Own Clues

Follow the clue guidelines below to create Mystery Data games like the ones you played. Write two different solutions to your game on a separate sheet of paper. Then, exchange your games with your classmates. Remember, there are likely to be multiple correct solutions.

Clues should provide the following information:

- type of data; for example, scores, hours, or prices
- number of values in the data set (Choose a number between 5 and 10.)
- the mean of the values
- at least one of the following: median, mode, range, lowest value, or highest value

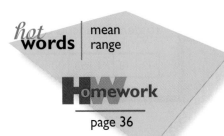

hot **words** | mean range

Homework

page 36

4 Top Teams

Stem-and-leaf plots are useful tools for representing and analyzing data. You can find out a lot by examining the shape of the data. You will use stem-and-leaf plots to compare sports teams.

Interpret Stem-and-Leaf Plots

What types of information can you get from stem-and-leaf plots?

Two teams of students timed how long they could balance on one foot with their eyes closed and arms extended overhead. Follow these steps to compare the teams.

1 On graph paper, make a stem-and-leaf plot for Team B. Be sure to put the numbers in order from least to greatest.

2 Find the median for each team.

3 How are the shapes of the two stem-and-leaf plots different? What does that tell you about the differences between the teams?

4 Which team do you think is better at balancing? Use data to support your conclusions.

Balancing Data

Team A: This stem-and-leaf plot shows how many seconds each student balanced on one foot.

Team B: Here are the numbers of seconds each student balanced on one foot.

33	18	41	26	30	35	38	49	27	46	47	29	36	44	45

Stems	Leaves
0	
1	9
2	2 4 5 6 6 6 8 9
3	0 3 5 8 9
4	
5	
6	9

Note:
2|4 means 24 seconds

Compare Stem-and-Leaf Plots

Your challenge is to figure out which of two mystery baseball teams is better at hitting home runs.

1 Use the data below to make a stem-and-leaf plot for each team.

2 Find the following for each team.

    a. least value   **b.** greatest value   **c.** range

    d. median   **e.** mean

3 What are the differences between the two teams? Write at least six comparison statements.

4 Which team do you think is better overall at hitting home runs? Why?

> **How can you use stem-and-leaf plots to make comparisons?**

Home-Run Data

The lists show the numbers of home runs hit by players on each team in a recent season. For example, a 6 means that one player hit 6 home runs during the season.

Team X

6	30	23	28	23	34	12	4	5	10	10	2	11	5	3

Team Y

0	42	22	5	19	2	2	9	5	14	3	6	5	6	5

Write Data-Based Statements

Take on the role of a professional baseball player for Team X or Team Y. An interviewer asks you these questions:

1 You have the honor of hitting the *median* number of home runs. How do you compare with the other players?

2 How does the best home-run hitter on your team compare with the other players?

3 Is your team better at hitting home runs than the other team? Give me the data to prove it.

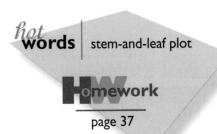

hot **words** | stem-and-leaf plot

Homework

page 37

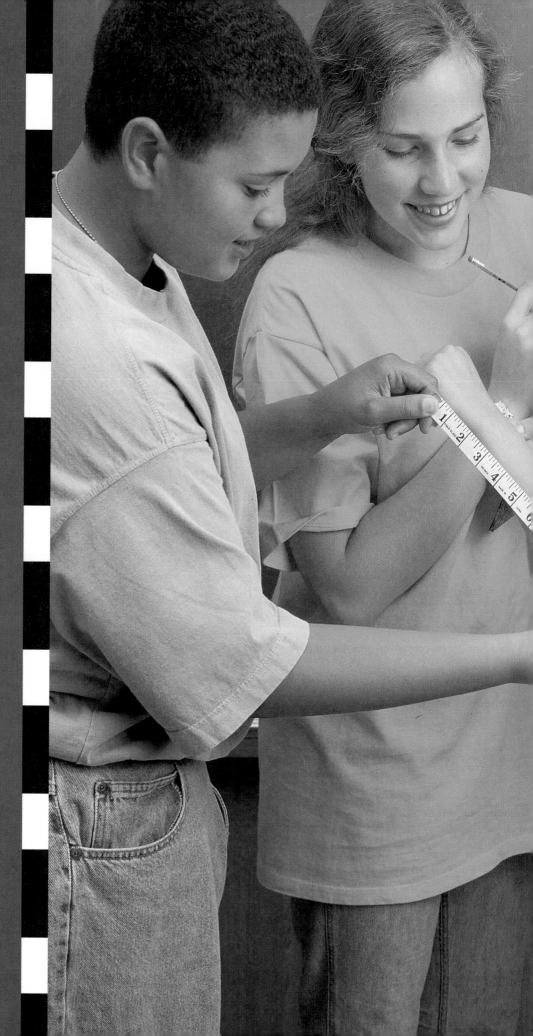

PHASE TWO

In this phase, you will investigate questions like, "Can you use foot measurements to make predictions about students' heights?" You will collect data and make scatter plots to see if there is a relationship between variables. You will also learn about different types of correlations.

Data, Scatter Plots, & Correlations

WHAT'S THE MATH?

Investigations in this section focus on:

STATISTICS: DATA COLLECTION and REPRESENTATION

- Collecting and organizing data
- Making back-to-back stem-and-leaf plots
- Making scatter plots to represent paired data

STATISTICS: DATA ANALYSIS

- Analyzing the relationship between pairs of data
- Identifying positive and negative correlations
- Sketching and interpreting lines of best fit
- Distinguishing between correlation, and cause and effect
- Making conclusions that are based on data

MEASUREMENT

- Estimating and measuring

MathScape Online
mathscape3.com/self_check_quiz

5 Comparing Sizes

Back-to-back stem-and-leaf plots are useful for making comparisons. You will collect measurement data about your body and plot it with data from your classmates. Then you will make comparisons and write your conclusions.

Collect Measurement Data

How can you make accurate measurements so that your data is useful for making comparisons?

When you collect data to compare sizes, it's important to make careful, accurate measurements. Sloppy measurements could throw off the class data and lead to false conclusions.

1 Starting with Line A, estimate its length to the nearest half centimeter and record your estimate. Then measure the line to the nearest half centimeter and record your measurement. Repeat this process with the other lines. Did your estimates improve?

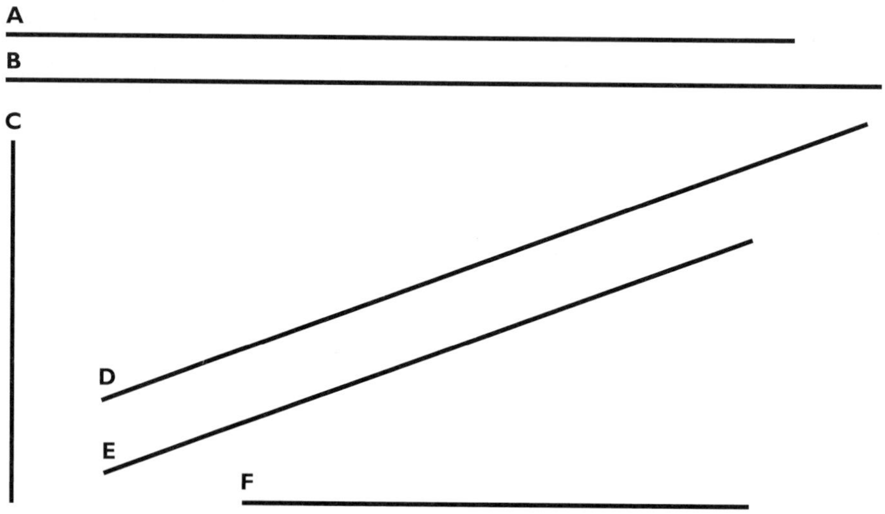

A ————————————————————

B ————————————————————

C
D
E
F ————————————

2 Your teacher will provide a Measurement Survey listing the body parts you should measure. Remember to measure to the nearest half centimeter.

Make Back-to-Back Stem-and-Leaf Plots

Now that your class has collected lots of data, it's time to make comparisons. Which are longer, students' feet or their forearms?

1 On graph paper, make a back-to-back stem-and-leaf plot for the class's foot and forearm data. Before you begin, make sure to put each data set in order from least to greatest values.

2 How do the shapes of the data compare for feet and forearms?

3 Find the mean, median, and range for lengths of feet and forearms.

4 In your class, which are longer, feet or forearms? Are they the same size? Use data to support your conclusions.

> **How can you use back-to-back stem-and-leaf plots to make comparisons?**

A Sample Back-to-Back Stem-and-Leaf Plot

In this type of plot, the stems are in the center of the plot.

Data from 23 Students in Ms. Rodriguez's Class

Foot Lengths (cm)		Stems	Forearm Lengths (cm)
		1	
	8	•	
5.5 5 4 4 4 2 2 0		2	0 0 2 2 4 4 5 5 5 5 5.5
9 9 8 8 8 7 7 6 6 6 6 6		•	6 6 6 6.5 7 8 8
	4 1.5	3	0
		•	6 6 7
		4	5
Leaves		**Stems**	**Leaves**

0 | 2 | represents 20 cm

| 2 | 0 represents 20 cm

Write Comparison Statements

When you make comparisons, it's important to use specific examples from the data.

- In your class data, how do the lengths of students' feet and forearms compare? Write at least 5 data-based comparisons.

- How does the back-to-back stem-and-leaf plot for your class compare with the sample one? Would your conclusions be the same or different for the two classes? Why?

hot **words** | stem-and-leaf plot
median

page 38

6 Is There a Relationship?

INTRODUCING
SCATTER PLOTS

Scatter plots are useful for showing the relationship between two variables, such as heights and weights. They are also called scattergrams and scatter diagrams. You will analyze a scatter plot and then create your own using data on mother animals and their babies.

Interpret a Scatter Plot

How can you identify points on a scatter plot?

Do students with longer feet tend to have longer forearms? You will interpret a scatter plot to see if there is a relationship between the two measurements.

1 Five of the points on the scatter plot have been labeled, *A, B, C, D, E.* Which of the following students does each point represent?

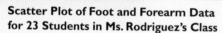
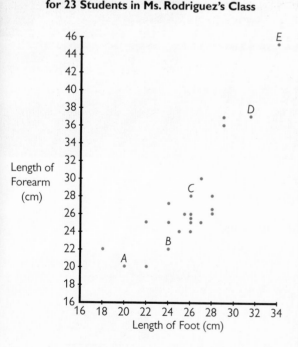

Scatter Plot of Foot and Forearm Data for 23 Students in Ms. Rodriguez's Class

Length of Forearm (cm)

Length of Foot (cm)

Simon: "I have the longest forearm and foot in the class."

Tanya: "My foot is the same length as my forearm."

Heather: "My forearm is about 5 centimeters longer than my foot."

Miguel: "My foot is the same length as those of four other students, but my forearm is longer than theirs."

2 The extra labeled point is for David's data. Write a description for him.

3 Based on the scatter plot, what do you think the relationship is between the lengths of students' feet and forearms? Why?

Make a Scatter Plot

Is there a relationship between the lengths of mother animals and their babies? Your teacher will give you a table of animal data. Use this data to investigate the following.

1 Choose two variables from the table. Make a scatter plot to show the relationship between the two variables.

2 Analyze the scatter plot. Do the points suggest a line? What is the relationship between the two variables? Why?

How can you make a scatter plot to show the relationship between pairs of data?

Guidelines for Making Scatter Plots

To make a scatter plot, you need pairs of data. The data can be number counts, such as the number of free throws made, or measurement data, such as heights. Scatter plots cannot be used for category data, such as colors or foods.

- The axes do not need to begin at zero. The two axes can have different scales. Choose scales that makes sense for your data.

- The intervals between the numbers on a scale need to be the same size. For example, 1, 2, 3, 4, or 2, 4, 6, or 10, 20, 30.

- If there is more than one pair of data for one point on a grid, you can use a number instead of a point. For example, if three pairs of data have the values 5 ft and 90 lbs, then use the number 3 instead of a point.

Write About Points on a Scatter Plot

Label four of the points on your animal scatter plot with the letters, *A, B, C, D*. Write a description of each point similar to the descriptions on the opposite page.

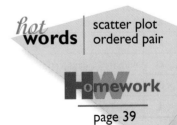

hot **words** | scatter plot
ordered pair

Homework

page 39

What Type of Relationship Is It

Scatter plots can help you see relationships in data. First, you will look at pairs of variables to see how their values are related, then you will make some scatter plots and look for correlations. You will see how a line of best fit can help you figure out if a correlation is weak or strong.

Discuss Relationships Between Variables

What types of relationships can two variables have?

Pairs of variables can be related in different ways:

- positive correlation: one variable increases when the other increases

- negative correlation: one variable increases when the other decreases

- no correlation

Look over the pairs of variables in the table below and answer these questions.

1 Which of the three relationships do you think each pair is most likely to have? Why?

2 Come up with your own examples for each type of relationship.

Pair	Variable 1	Variable 2
A	amount of trash a family recycles	amount of trash a family sends to the dump
B	number of rainy days per year	number of dry days per year
C	length of an object in inches	length of an object in centimeters
D	age	length of hair
E	money spent on bags of pretzels	number of bags purchased
F	weight of package	cost of postage
G	number of days a student is absent from school	number of days a student attends school
H	number of teenagers in house	amount of time phone is in use in house
I	number of candy bars eaten in a typical week	number of cavities

Identify Different Correlations

Is there a correlation between the popularity of a food and the amount of fat, sugars, or sodium in a serving? Make a scatter plot to find out.

1 Use your class data to make a scatter plot.

2 What type of correlation does the scatter plot show?

3 Draw a line of best fit for the food scatter plot. If it doesn't make sense to draw this line for your scatter plot, explain why.

4 How close are the points to the line of best fit? What does this tell you about the relationship between the two variables?

> **How can you determine what type of correlation a scatter plot shows?**

Different Types of Correlations

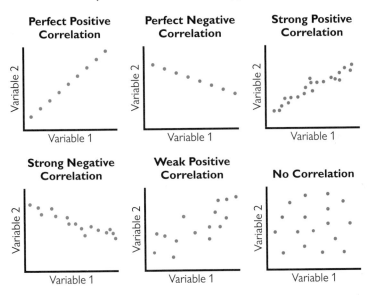

These scatter plots show different types of correlations.

Drawing a **line of best fit** can help you figure out the relationship between two variables in a scatter plot. Use a ruler and pencil to draw a line that follows the trend of the data. In some scatter plots, the line will connect many of the points. In others, you will need to draw a line so that about half the points are above the line and half are below. If most of the points are on the line or close to the line, that shows that there is a strong correlation between the variables. See page 22 for an example.

hot **words** | correlation
line of best fit

page 40

8 The Mysterious Footprint

APPLYING
SCATTER PLOTS
AND CORRELATIONS

This is your chance to be a detective. You will use your knowledge of scatter plots and correlations to investigate a mystery.

Make Reasonable Estimates

How can you use a line of best fit to make predictions?

On the day of the Baker Middle School bake sale, a large pan of brownies was stolen from the cooking room. Fortunately, the floor was coated with flour, and the thief left a floury footprint 27 cm long. Your challenge is to use this clue and your knowledge of correlations to make a reasonable estimate about the thief's height.

1 Analyze the scatter plot on your handout. What type of relationship does it show between heights and foot lengths?

2 What would you estimate the thief's height to be for each of the following foot lengths?

    a. 27 cm   **b.** 31 cm   **c.** 32 cm   **d.** 23 cm

How to Make an Estimate

If there is a correlation between the two variables, you can use the line of best fit to make estimates. For example, if a footprint is 33 cm long, you could estimate that the thief's height is about 180 cm. The stronger the correlation, the more accurate the estimates are likely to be.

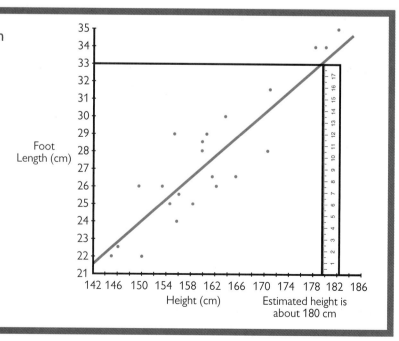

Investigate Class Data

There were two more clues left at the scene of the Baker Middle School crime: a wristwatch and a chocolate-covered, long-sleeved shirt (see below). What estimates can you make about the thief's measurements based on these clues and your class data?

How can you make and test hypotheses about which pairs of measurement data have strong correlations?

Watch: 16 cm

Footprint: 27 cm

Shirt Sleeve from shoulder to wrist: 51 cm
Neck circumference: 32 cm

Planning and Organizing

1 Choose a variable from one of the clues (foot, wrist, neck, or arm measurements). Then choose a different variable (height, forearm, calf, or ankle) from your class measurement survey.

2 Make a hypothesis. What kind of relationship do you think that these variables have? Why?

3 Make a table to organize the pairs of data.

Representing and Analyzing the Data

4 Make a scatter plot of the data. How did you set up the axes?

5 Draw a line of best fit. If it doesn't make sense to draw one, explain why not.

6 How would you describe the points? What trends do you see?

Making Conclusions and Estimates

7 What type of correlation does your scatter plot show: positive, negative, or no correlation? Why? Give examples of at least four points that have this relationship.

8 Do your conclusions support your hypothesis? Why?

9 Write an explanation of how to use a line of best fit to make an estimate about the thief's measurements. If your scatter plot shows a correlation, make an estimate. For example, if you plotted data on wrists and necks, estimate the thief's neck measurement. Draw lines on your scatter plot to show how you made the estimate.

hot **words** | scatter plot correlation

Homework

page 41

PHASE THREE

To: Students

    You will take on the role of a member of a music group that is about to go on tour. You can choose to be a musician, singer, sound engineer, or manager. You will use the mathematics of probability, combinations, and permutations to solve problems that arise on the tour.

In this phase, you will be playing games of chance and figuring out the probabilities of events in the games. You will learn some useful techniques, such as making tree diagrams and using the Permutation Theorem. These tools will help you to find all the different combinations of a set of objects.

Probability, Combinations, and Permutations

WHAT'S THE MATH?

Investigations in this section focus on:

PROBABILITY and STATISTICS

- Collecting and analyzing data
- Determining theoretical probabilities
- Determining experimental probabilities

NUMBER

- Figuring out different combinations of objects
- Figuring out different permutations of objects
- Using systematic listings
- Making tree diagrams
- Using the Permutation Theorem

MathScape Online
mathscape3.com/self_check_quiz

On Tour

EXPLORING
PROBABILITY AND
COMBINATIONS

Have you ever wanted to go on tour with a band? You will play a game of chance to collect data on different events that can happen while on tour. Then you will analyze your data.

Play a Game of Chance

How can you analyze data to find out about the probabilities of events?

In the Concert Tour Game, you will use a number cube, spinner, and coin to randomly determine what happens when your group goes on tour. Keep track of your data in the table on the handout.

Analyze the game rules and the data to answer these questions:

1 How many times did you get each of the following:

  a. ideal location?   **b.** ideal audience size?

  c. ideal audience reaction?

2 Based on your data, what is the experimental probability of each of the above events?

3 What is the theoretical probability of each of the above events?

4 How many times did all three of your ideal events come up on one turn?

Probabilities

Theoretical probabilities are found by analyzing a situation, such as looking at game pieces or game rules. You can use the following ratio:

$$\frac{\text{number of favorable outcomes}}{\text{total number of possible outcomes}}$$

Experimental probabilities are based on data collected by conducting experiments, playing games, and researching statistics in books, newspapers, and magazines. You can use the following ratio:

$$\frac{\text{number of times a particular outcome occurred}}{\text{total number of tries or turns}}$$

List All the Possible Combinations

How can you use systematic listing to figure out all the different combinations?

In order to figure out the theoretical probability of getting all three of your ideal events on one turn, you need to know how many possible combinations of events there are.

1 Start by simplifying the problem. List all the possible combinations for just two events: locations and audience reactions. For example: 1H (Boston, Happy), 1T (Boston, Thumbs Down). The order of the events doesn't matter, since 1H is the same combination as H1.

2 How many different combinations did you find? How can you make sure you found them all?

3 List all the possible combinations of the three events. How many combinations did you get?

4 What is the theoretical probability of all three of your ideal events coming up on one turn?

Improve the Game

The game designers would like your help in changing the Concert Tour Game. They want to use the three spinners below instead of a number cube, a coin, and a spinner.

1 How would you label each of the spinner parts? Brainstorm ideas. Then copy the spinners and fill in your labels.

2 What are your three ideal events in this new game?

3 What is the theoretical probability of getting all three of your ideal events on one turn? Explain how you figured it out.

Concert locations

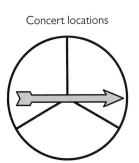

Transportation: How will you get there?

What will you wear?

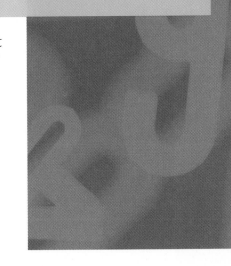

hot words | experimental probability
theoretical probability

**omework**

page 42

10 Lunch Specials

Is a restaurant's special as good a deal as it claims? You can use tree diagrams to help answer this question.

Make a Tree Diagram

How can you use tree diagrams to figure out all the possible combinations of things?

When your music group is on tour, you are on a tight budget. That's why you're always on the lookout for restaurants with special deals. Figure out whether the special shown below is as good as it is claimed to be.

1 Make a tree diagram like the one on your handout to figure out all the possible lunch combinations.

2 Circle the path that shows the combination that you like best. How much money would you save by ordering the special instead of buying each item separately?

3 How many different combinations are there? Count the bottom branches to find out.

4 Is the advertisement's claim true? Why or why not?

The Unique Diner's Lunch Special

Choose one main dish, one side dish, and one beverage for just $2.50. There are over 21 unique combinations, and each one saves you $1 or more!

Main Dish Choice	Side Dish Choices	Beverage Choices
Hamburger $2.25	French Fries $0.99	Milk $0.65
Burrito $1.99	Salad $1.25	Juice $0.75
Pizza $1.80	Corn $0.55	

Compare the Numbers of Combinations

The table shown here depicts how many different lunch choices are offered at six different restaurants. Customers can choose one main course, one beverage, and one side dish for a special price.

Restaurant	Number of Main Dishes	Number of Side Dishes	Number of Beverages	How many combinations?
A	3	2	2	?
B	3	3	2	?
C	4	3	2	?
D	4	3	3	?
E	4	3	4	?
F	5	3	4	?

How does adding more choices affect the total number of different combinations?

1 Find the number of different combinations for each restaurant.

Tip: Make a tree diagram for Restaurant A. Either decide what the dishes will be (tacos, spaghetti, hot dogs, and so on) or use letters and numbers to stand for each dish. For example, Restaurant A has three main dishes (M1, M2, M3), two side dishes (S1, S2), and two beverages (B1, B2).

2 Keep track of the number of combinations for each restaurant by making a table as shown. What patterns do you see?

Create Dinner Specials

Create and investigate dinner menus by following these steps:

1 Make a dinner menu that has 4 main dishes, 2 side dishes, 2 beverages, and 2 desserts. Exchange menus with a classmate.

2 For your classmate's menu, make a tree diagram to figure out all the possible combinations consisting of 1 main dish, 1 beverage, 1 side dish, and 1 dessert.

3 Use the Fundamental Counting Principle to check that you found all the combinations. This principle states that if the first category has m options and the second category has n options, and if you choose one option from each category, then the number of combinations is $m \cdot n$.

hot **words** | combination
tree diagram

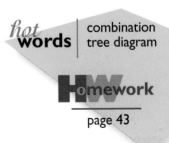

Homework

page 43

11 The Battle of the Bands

INTRODUCING PERMUTATIONS

Your music group is going to compete in a "Battle of the Bands." Do you want to perform first, second, third, or last? The order will be chosen randomly. You will figure out how many different orderings there are and calculate the probability of getting the one you want.

Experiment with Permutations

How many different orderings can you get if you randomly choose the order of four things?

Do an experiment to find out how many times your music group will be randomly chosen to perform first, second, third, or fourth. Follow the directions on the handout What's the Order?

After you do the experiment, answer these questions.

1 How many different orderings or permutations did you get?

2 Did you get the same ordering more than once?

3 Do you think you got all the possible orderings in your experiment? Why or why not?

4 Would you prefer to perform first, second, third, or fourth? How many times did you get your preference?

Permutations

A **permutation** is an arrangement of names, letters, or objects in a particular order. For example, imagine that your group is going to perform in three different cities: Atlanta, Boston, and Chicago. You get to choose the order in which you visit the cities. Here are three different permutations.

1. Atlanta ⟶ Boston ⟶ Chicago

2. Boston ⟶ Chicago ⟶ Atlanta

3. Chicago ⟶ Atlanta ⟶ Boston

What other permutations can you find?

Find All the Permutations

Find all the possible orderings or permutations for the four music groups by following these steps.

1 Copy the tree diagram below and complete it for the four bands. Each letter stands for a different band. Decide which letter represents your band.

2 How many different permutations are there?

3 Find the theoretical probability of randomly choosing each of the following orderings.

  a. Your band is first.

  b. Your band is last.

  c. Your band is second or third.

  d. Band B is first, and Band C is last.

  e. D, C, B, A

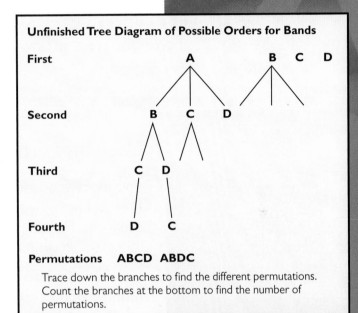

How can you use a tree diagram to find all permutations of four objects?

Unfinished Tree Diagram of Possible Orders for Bands

First   A   B   C   D

Second   B   C   D

Third   C   D

Fourth   D   C

Permutations   **ABCD**   **ABDC**

Trace down the branches to find the different permutations. Count the branches at the bottom to find the number of permutations.

Create Word Puzzles with Permutations

You can use what you know about permutations to write and solve word puzzles.

1 Here are permutations of the letters in two words:

  a. trposs   **b.** rigsne

What are the words?

2 Write your own word puzzle. Choose a word that has 5 to 10 letters. Try out at least three different permutations of the letters. Choose one of them for your puzzle.

3 Exchange puzzles with a classmate.

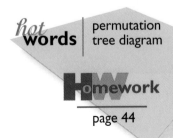

hot **words** | permutation
tree diagram

Homework

page 44

12 The Demo CD

Suppose a large number of bands are going to compete in a contest. How many different orderings of the bands are there? You will see how the Permutation Theorem can help answer this question.

Use the Permutation Theorem

How can you use the Permutation Theorem to figure out the number of possible orderings for a large number of objects?

Five more bands want to enter the Battle of the Bands contest. It would be difficult to make a tree diagram to show all of the orderings for nine bands. Another way to solve the problem is to use the Permutation Theorem.

Use the Permutation Theorem and a calculator to figure out the number of different orderings possible for each of the following numbers of bands:

a. 2	b. 3	c. 4	d. 5
e. 6	f. 7	g. 8	h. 9

The Permutation Theorem

The **Permutation Theorem** says that there are $n!$ (n factorial) possible permutations of n objects. (Each object can be used only once.)

For example, for 4 bands, there are 4! (or 4 factorial) permutations. To calculate 4!, multiply $4 \times 3 \times 2 \times 1$.

Number of bands that can go first: 4
Number that can go second: 3

(After one is picked to go first, you are choosing from 3 bands to go second.)

Number that can go third: 2
Number that can go fourth: 1

$4 \times 3 \times 2 \times 1 = 24$ possible permutations or orderings

Investigate Different Situations

1 The band is getting offers to perform all over the country. On each tour they arrange, they will play at 6–10 different locations. There may be 2 or 3 different audience reactions. To allow for each set of possibilities, a revised Concert Tour Game is played with 3 spinners, each divided into equal parts. Make a table like the one here and complete the missing information. Explain how you figured it out.

How can you apply what you know about combinations, permutations, and probability?

Game Version	Number of Locations	Number of Audience Sizes	Number of Audience Reactions	Number of Possible Combinations
Original	6	3	2	?
Version B	8	2	2	?
Version C	9	4	3	?
Version D	?	?	?	90

2 For each version of the game, what is the probability that you would get your ideal location, audience size, and audience reaction on one turn? Rank the games in order from least to greatest probability. Explain how you figured this out.

3 Your band gets a contract to make a demo CD. Which of your three favorite songs should go first, second, or third? Write down the song names, then make a tree diagram to show all the possible permutations, or orderings, of these songs.

4 Your group has 11 songs to put on the demo CD. Use a calculator to figure out how many possible orderings there are for 11 songs. Explain how you figured it out.

Write About the Mathematics

Write a letter to a student who will be doing this unit next year. Describe the investigations and the mathematics you learned.

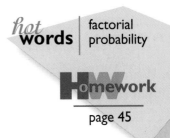

hot **words** | factorial probability

Homework
page 45

Sports Opinions and Facts

Applying Skills

In Mr. Jackson's class, 25 students rated how much they enjoyed playing soccer.

1 (Not at all)	ⱵⱵ Ⅲ
2	ⅠⅠⅠⅠ
3	ⱵⱵ Ⅰ
4	ⅠⅠ
5 (Very much)	ⱵⱵ

1. How many students chose each rating?

2. What percentage of students in the class gave a rating of 1? of 3? of 5?

3. Add up all the ratings to find the total.

4. What is the class average, or mean rating?

Extending Concepts

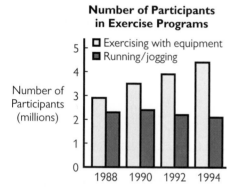

Number of Participants in Exercise Programs

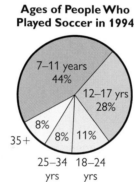

Ages of People Who Played Soccer in 1994

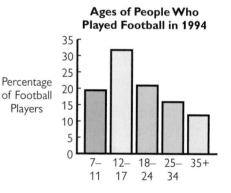

Ages of People Who Played Football in 1994

Give specific data to support your answer to each question below.

5. Of the people who played football in 1994, what percentage was over 25 years of age?

6. Of the people who played soccer in 1994, were more of them between 12 and 17 years of age or over 18?

7. Which group is generally older, the people who played soccer in 1994 or the people who played football in 1994?

8. In 1990 which was more popular, jogging or exercising with equipment?

9. During the period 1988–1994, did jogging gain in popularity? Did exercising with equipment gain in popularity?

Writing

10. Write two questions of your own that can be answered using the graphs above and two that cannot be answered using these graphs.

Who's the Best?

Applying Skills

Find the mean and median for each data set.

1. 12, 19, 16, 30, 11

2. 2, 8, 11, 7, 17, 5, 5, 1

3. 1.5, 6.2, 9.9, 2.8

Find the range for each data set.

4. 2, 5, 7, 14, 9, 12

5. 1, 26, 87, 2, 90, 55

6. 5.0, 4.8, 5.0, 5.1, 4.9

7. In nine basketball games, Marta scored 11, 18, 20, 15, 18, 8, 6, 13, and 26 points. Find her mean, median, mode, high score, and range.

8. In six gymnastics competitions, Jacy scored 9.1, 9.8, 9.5, 8.8, 9.1, and 9.9 points. Find his mean, median, mode, high score, and range.

The table shows statistical measures for three basketball players. The statistics are based on the data for nine games.

	Mean	Median	Mode	High Score	Range
Player A	16	16	15	18	4
Player B	9	10	6	14	9
Player C	18	18	9	28	22

Rank the players from best to third best by each of the following:

9. mean

10. median

11. mode

12. high score

Extending Concepts

13. Which player (A, B, or C) was the most consistent in the number of points scored? How can you tell?

14. Bill claimed that Player A was the best player. Maria claimed that Player C was the best player. Use the data to write an argument in support of each person's claim.

15. Which player (A, B, or C) would you choose to put in a game if you are playing an excellent team? Only an outstanding performance will give you a chance of winning. Give reasons for your choice.

Making Connections

16. These are the approximate speeds (in miles per second) at which the planets orbit the sun. Find the mean, median, mode, and range of the speeds.

Mercury	30
Venus	22
Earth	19
Mars	15
Jupiter	8
Saturn	6
Uranus	4
Neptune	3
Pluto	3

Mystery Data

Applying Skills

For each data set, find the mean, median, mode, and range.

1. 3, 5, 9, 14, 3, 2, 10, 10, 3, 15, 8

2. 2.5, 3.1, 4.0, 7.2, 1.7, 2.5

Create a data set that matches each set of clues.

3. Tyler played 5 basketball games.

The mean was 10 points.

The median was 12 points.

What might each of his scores be?

4. Desrie played 6 basketball games.

The mean was 18 points.

The range was 8 points.

What might each of her scores be?

5. Six students kept track of how much time they spent reading in a week.

The mean time was 6 hours.

The mode was 7 hours.

What might each of their times be?

6. Francisco bought 5 books.

The mean price was $23.80.

The most expensive book cost $30.

What might each of the books cost?

Extending Concepts

7. a. When averages are used to make comparisons, everyone wants to be above average, and no one wants to be below average. Make a data set that has seven values, a mean of 50, and only one value that is below the mean.

b. Do you think it is possible to create a data set that has seven values, a median of 50, and only one value below the median? Explain your thinking.

8. Create a data set that has five values, a median of 30 and:

a. a mean that is greater than its median.

b. a median that is greater than its mean.

Explain how you created the data sets.

Writing

9. Answer the letter to Dr. Math.

> Dear Dr. Math:
>
> I'm confused. We've been studying means, medians, and modes. I'm always getting them confused because they all begin with the letter *m*. What's the difference between these things?
>
> Mark M. Morrison

Top Teams

Applying Skills

The stem-and-leaf plot below shows the number of home runs for each player on a baseball team during one season.

```
Stems  Leaves
  0 | 2 3 5 5 7 8 8 8
  1 | 1 4 6
  2 | 0 2 5
  3 | 1
       3 | 1 = 31 home runs
```

1. What is the greatest number of home runs? the least?

2. What is the median number of home runs?

3. What is the mean number of home runs?

4. What is the range?

5. How many players got 8 home runs? fewer than 10? more than 19?

The following are the number of home runs scored by the players of a second baseball team during the same season.

6, 17, 12, 11, 6, 21, 5, 12, 9, 14, 24, 4, 25, 14, 18

6. Make a stem-and-leaf plot for this second team.

7. Use your plot to find the median, highest and lowest values, and range.

Extending Concepts

The stem-and-leaf plots show the average number of points per game for basketball players of the New York Knicks and the Orlando Magic during one playing season.

```
  New York              Orlando
Stems  Leaves         Stems  Leaves
  0 | 2 3 4 4 5 6        0 | 2 3 4 4 5 5 6 9
  1 | 0 1 1 3 4 5        1 | 3 5 8
  2 | 3                  2 | 2 7
     2 | 3 = 23 points      1 | 3 = 13 points
```

8. Find each team's mean, median, and range.

9. Write a paragraph comparing the two teams. Use your answers to item **8** and compare the shapes of the two plots.

Making Connections

10. The table shows the average life span of 11 mammal species. Make a stem-and-leaf plot to represent this data. Use your plot to find the median life span.

Animal	Life Span (years)
Elk	15
Deer	8
Tiger	16
Possum	1
Fox	7
Camel	12
Mouse	3
Elephant	35
Wolf	5
Monkey	15
Grizzly Bear	25

Comparing Sizes

Applying Skills

The back-to-back stem-and-leaf plot shows the heights of 15 college football players and of 15 college basketball players.

Heights of Football Players (in.)		Heights of Basketball Players (in.)
	6	
9	•	8
5 4 3 3 2 2 1 1 0	7	3 5
8 7 7 6 6	•	6 6 6 7 8 9
	8	0 0 1 4
	•	6 7

0 | 7 | = 70 inches | 7 | 3 = 73 inches

1. What is the height of the tallest football player?

2. How many basketball players are over 82 inches tall?

3. Find the mean, median, and range of the basketball players' heights.

4. Find the mean, median, and range of the football players' heights.

5. In general, are the football players or the basketball players taller?

Extending Concepts

6. How do the heights of the football players compare with the heights of the basketball players? Use the back-to-back stem-and-leaf plot to write five data-based comparison statements. Be sure to include statements comparing the shapes of the two data sets.

Making Connections

The table shows the 1995 life expectancy for females in ten European countries.

Country	Life Expectancy
Sweden	81
Portugal	79
Germany	80
Hungary	76
United Kingdom	80
France	82
Poland	77
Italy	80
Belgium	81
Denmark	79

7. Find the mean, median, and range of the data set.

8. What additional data could you look up in order to make a back-to-back stem-and-leaf plot with the data in the above table? (The data in the table would be one side on the plot; your new data would be on the other side.) What might you be able to find out from your back-to-back stem-and-leaf plot?

Is There a Relationship?

Applying Skills

Use the scatter plot for items **1** and **2**.

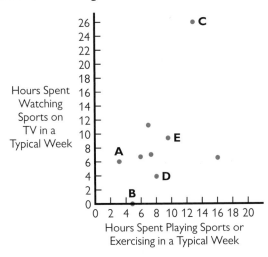

Hours Spent Watching Sports on TV in a Typical Week

Hours Spent Playing Sports or Exercising in a Typical Week

1. Which of the students below does each of the points *A, B, C, D, E* represent?

Matthew: "Each week I spend more than 12 hours playing sports. I also spend more than 12 hours watching sports on TV."

Jong: "I spend the same amount of time playing sports as watching them on TV."

Lyn: "I spend twice as much time watching sports on TV as playing them."

LaToya: "I spend time playing sports, but I don't watch them on TV."

2. One of the points doesn't have a match. Write a description for that point.

Extending Concepts

3. According to the scatter plot, which statement best describes the relationship between time spent playing sports and time spent watching sports on TV?

a. As one variable increases, the other increases.

b. As one variable increases, the other decreases.

c. There is no relationship between the variables.

4. What conclusions would you draw from the graph about the relationship between the number of hours a student plays sports and the number of hours he or she watches sports on TV?

Making Connections

The table shows the latitude and a typical minimum temperature in January for each of eight U.S. cities.

City	Latitude (°N)	Minimum Temperature in January (°F)
Anchorage	61	8
Miami	26	59
Helena	47	10
Buffalo	43	17
Reno	39	21
Memphis	35	31
Houston	30	40
Chicago	42	13

5. Make a scatter plot showing the relationship between the two variables.

6. What trend does this set of data show?

7. What can you say about the relationship between the two variables?

What Type of Relationship Is It?

Applying Skills

For each of the following, determine whether or not you can make a scatter plot. If you can, draw and label the axes. Describe what type of relationship there would be. If you cannot make a scatter plot, explain why not.

1. lengths and weights of cats

2. students' favorite sports and favorite music groups

3. heights of 8th graders

4. students' ages and the amount of allowance they receive

Tell whether each scatter plot shows positive, negative, or no correlation.

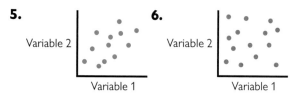

5. Variable 2 / Variable 1

6. Variable 2 / Variable 1

7. Variable 2 / Variable 1

Extending Concepts

8. Which plot from items 5–7 shows the strongest correlation? How can you tell?

Writing

9. Answer the letter to Dr. Math.

Dear Dr Math,

I'm having trouble figuring out how to read a scatter plot. There are so many little dots all over the place. How can you tell if there is a correlation? Please give me some examples and draw some pictures to help me figure this out.

Dottie

The Mysterious Footprint

Applying Skills

This scatter plot shows ages and weights of some monkeys at a zoo.

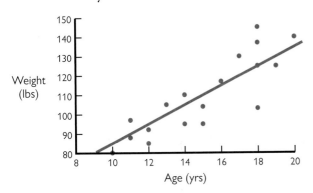

1. Use the scatter plot to estimate the weight of monkeys of the following ages:

   a. 13   **b.** 16   **c.** 19

2. Use the scatter plot to estimate the age of a monkey whose weight is the following:

   a. 90 lb   **b.** 110 lb   **c.** 130 lb

Extending Concepts

3. Give an example of a pair of variables that are likely to have no correlation and a pair that are likely to have negative correlation.

4. Give an example of a pair of variables that are likely to have a strong positive correlation.

Writing

5. Answer the letter to Dr. Math.

> Dear Dr. Math,
>
> A friend of mine told me that I could use scatter plots to make predictions. Then he said I couldn't make predictions from just any scatter plot. He also said certain scatter plots are better for making predictions than others. I don't know what he was talking about. Can you help me sort this out?
>
> Scattered in Scranton

On Tour

Applying Skills

On each turn of a game, players spin the two spinners shown and flip a coin. They get a point for a 1, a point for blue, a point for tails, and a bonus point for getting all three in one turn.

Pei's results for 12 turns are shown here.

Spinner 1	Spinner 2	Coin
2	blue	H
3	blue	H
4	red	T
2	pink	H
1	red	T
3	blue	H
3	red	T
4	pink	H
1	blue	T
4	pink	T
1	blue	H
4	red	H

1. How many times did Pei get:

   a. a 1?   **b.** blue?   **c.** tails?

2. Based on Pei's data, what is the experimental probability of getting each of the above events?

3. What is the theoretical probability of getting each of the above events?

4. How many times did Pei get 1, blue, and tails on one turn? What is the experimental probability of getting all three events on one turn?

Extending Concepts

5. List all the possible combinations for the two spinners. For example: 1B (1, blue) is one combination. How many combinations did you find?

6. List all possible combinations for the two spinners and the coin. 1BH (1, blue, heads) is one combination. How many did you find?

7. Find the theoretical probability of getting 1, blue, and tails in one turn.

Making Connections

8. The Native American game of Totolospi was played by the Moki Indians of New Mexico. Players would throw three flat throwing sticks, which were plain on one side and colored on the other, and score points if all sticks fell the same way. List all the possible combinations when the sticks are thrown. For example, CCP is one combination (stick 1: colored, stick 2: colored, stick 3: plain). What is the theoretical probability that all three sticks land the same way?

Lunch Specials

Applying Skills

The tree diagram shows the possible combinations for a lunch special.

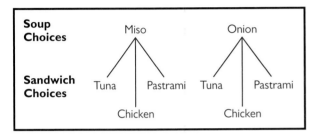

Soup Choices	Miso		Onion	
Sandwich Choices	Tuna	Pastrami	Tuna	Pastrami
		Chicken		Chicken

1. Write down each combination using initials.

2. How many different combinations are there?

Make a tree diagram to show the different lunch combinations. Find the total number of combinations if you can choose from the following. In each case, you can choose one item from each category.

3. 3 soups (tomato, leek, bean), 3 sandwiches (egg, tuna, cheese)

4. 2 sandwiches (cheese, turkey), 3 desserts (pie, ice cream, cake), 2 beverages (soda, juice)

Without making a tree diagram, find the number of lunch combinations if you can choose from the following. In each case, you can choose one item from each category.

5. 3 sandwiches, 4 beverages

6. 2 sandwiches, 3 soups, 5 beverages

7. 2 soups, 6 sandwiches, 4 desserts

8. 4 soups, 7 sandwiches, 8 desserts

Extending Concepts

9. Do you think the advertising claim below is true? Why or why not?

> **Unique Diner's**
> **$1.99 Breakfast Special!**
>
> Choose one item from each category: 4 types of pancake, 3 kinds of egg, 3 kinds of cereal, and 3 kinds of juice. You could eat breakfast here for 100 days and have a different combination each day!

Writing

10. Answer the letter to Dr. Math.

Dear Dr. Math:

I would like to offer 50 unique combinations for my lunch special at the Unique Diner. How many main dishes, side dishes, and/or beverages do I need? Please explain how you figured this out, so I can do it on my own.

The owner of the Unique Diner

The Battle of the Bands

Homework 11

Applying Skills

1. Copy and complete the tree diagram to find all possible permutations of the letters *X, Y, Z*. How many permutations are there?

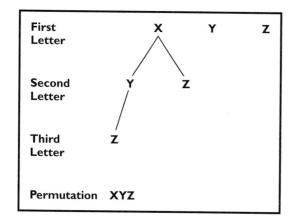

First Letter	X	Y	Z
Second Letter	Y	Z	
Third Letter	Z		
Permutation	XYZ		

Emma, Fiona, and Gina are three skaters in a competition. The order in which they skate will be picked randomly.

2. Make a list of all possible orderings (permutations) for the skaters.

3. How many permutations are there?

4. Find the theoretical probability of each of the following:

   a. Gina will skate first.

   b. Fiona will skate either first or last.

   c. Emma will skate last.

   d. The order will be Emma, Fiona, Gina.

Extending Concepts

5. Five bands (A, B, C, D, E) have offered to play at a party. Only two bands will be selected to play. The organizer will decide which band will play first and which band will play second by picking names from a bag.

   a. Make a tree diagram to show all possible orderings of 2 bands.

   b. How many possibilities are there?

   c. Find the theoretical probability that band C will play first, followed by band B.

Making Connections

6. a. Find all the possible permutations of the letters *a, i, l, r*. How many permutations are there?

   b. How many of the permutations in item **6a** represent English words? Which ones? (Note: The *aril* of a seed is its exterior covering. The *lira* is the currency in Italy, and the *rial* is the currency in Iran.)

The Demo CD

Applying Skills

Use a calculator to find each of the following:

1. 3! **2.** 5! **3.** 8!

4. 10! **5.** 12!

Use the Permutation Theorem and a calculator to find the number of possible orderings for the following number of skaters in a contest:

6. 4 **7.** 6 **8.** 9

9. 10 **10.** 13

11. Seven bands (A, B, C, D, E, F, G) are playing in a contest. The order in which they play will be selected randomly. How many different orderings are there?

Extending Concepts

Five TV game show contestants play a game of chance. Each player spins the spinner shown and rolls a number cube (with the numbers 1–6 on it).

12. Use a tree diagram to show all the possible combinations for the spinner and the number cube. How many combinations are there?

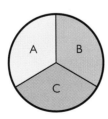

On each turn of the game show, points are scored as follows:

spin an A = 4 points

spin a B and roll an even number = 7 points

roll 6 and spin a C = 10 points

13. What is the probability that in a single turn a player will score the following:

a. 4 points? **b.** 7 points? **c.** 10 points?

The winner of the game show continues to round 2. In round 2, each of the letters R, O, M, and E is put in a bag. The player selects the letters randomly one at a time. If the letters are picked in the order R-O-M-E, the player wins a trip to Rome.

14. How many possible orderings are there for the letters R, O, M, E?

15. Find the theoretical probability that a participant will win round 2.

Writing

16. Answer the letter to Dr. Math.

> Dear Dr. Math,
>
> How can you tell if a problem is about finding combinations or permutations? What's the difference between these two things anyway? Please give me some examples.
>
> Combi Nation

Glencoe

This unit of MathScape: Seeing and Thinking Mathematically was developed by the Seeing and Thinking Mathematically project (STM), based at Education Development Center, Inc. (EDC), a non-profit educational research and development organization in Newton, MA. The STM project was supported, in part, by the National Science Foundation Grant No. 9054677. Opinions expressed are those of the authors and not necessarily those of the Foundation.

CREDITS: Photography: Chris Conroy • © M. Tcherevkoff/Image Bank: p. 2 • © Peter Zeray/Photonica: pp. 3TR, 24 • © Reza Estakhrian/Tony Stone Images: pp. 3TL, 4 • © Aric Crabb, courtesy of San Jose Lasers/American Basketball League: p. 10 • © T. Davis and W. Bilenduke/Tony Stone Images: p. 19.

Send all inquiries to:
Glencoe/McGraw-Hill
8787 Orion Place
Columbus, OH 43240-4027

ISBN: 0-07-866820-4

4 5 6 7 8 9 10 058 08 07